Subtraction

Alexander Gelman

Subtraction: Aspects of Essential Design

Editors: Baruch Gorkin, Kate Noël-Paton, Kaoru Sato
Research Coordinator: Yanitza Tavarez
Art Directors: Alexander Gelman, Kaoru Sato

 RotoVision

A RotoVision Book
Published and distributed by
RotoVision SA, Rue du Bugnon 7, CH-1299
Crans-Près-Céligny, Switzerland

RotoVision SA, Sales & Production Office
Sheridan House, 112/116A Western Road
Hove, East Sussex, BN3 1DD, UK

Tel: +44 (0) 1273 72 72 68
Fax: +44 (0) 1273 72 72 69
E-mail: sales@rotovision.com

Distributed to the trade in the United
States by Watson-Guptill Publications
1515 Broadway, New York, NY 10036

10 9 8 7 6 5 4 3 2 1

ISBN 2-88046-391-2

Book design by Design Machine, New York

Production and separations in Singapore
by ProVision Pte. Ltd.

Tel: +65 334 7720
Fax: +65 334 7721

I wish to thank all who helped make this book possible. Above all, the artists and designers that responded to my call. Mel Byars, for his energy and enthusiasm that brought the project to life. Paola Antonelli, for introducing me to Mel. Baruch Gorkin, for helping shape the book's concept. Boris Bencic, Jasper Bode, Michel Bozorg, Martin Ferran-Lee, Paul Jarvis, Andrew Logvin, Kimi Nagoshi, Barbro Ohlson, Shen Ru, Frederik Sward, Chigako Takeda, Takako Terunuma, Aia Urakawa, and Laetitia Wolff, for research assistance. David Boyer and Jeff Newlet, for participating in writing and editing. The Design Machine staff: Alissyn Brock, Pamela Korblak, Kaoru Sato, Toby Moore, Tanya Pramongkit, Yanitza Tavarez, and Amaya Taveras, for their support. Special thanks to Jonathan Barnbrook, Joshua Goodman, Michael Horsham, Tibor Kalman, Jeffrey Keyton, John Maeda, and James Wines.

To my parents Eugenia and Gregory Gelman

Subtraction is not an ism.
By Mel Byars

This book first and foremost is about excellence and the manifestations of clear thinking. However, just what qualities or circumstances foster or motivate excellence is a mystery to me. Certainly it is not about the time required to realize an idea in art, architecture, or design — not even in writing. Jackson Pollock, for example, completed a painting in a matter of hours. Lorenz Hart wrote one of his better-known tunes in a restaurant on a paper coffee cup. Voltaire penned *Candide* in seven days. And Brassaï captured a story in a split second through photography. Then, there are those, of course, who labor for months and years with no less admirable results. So excellence is not about time. And it is not about academic education nor high intelligence. The best and the worst human characteristics that foster excellence may arguably be obsession and ego.

Human beings have a need to communicate ideas, thoughts, feeling, emotions, instructions. Evidence reaches back very far. Artifacts have revealed that the advent of human expression through two- and three-dimensional images reaches back as far as 30 millennia and that stone tools appeared as long ago as two and a half million years. But advances in technology continuously extend the date further and further. Examples are being found in ideographs, pictographs, and petroglyphs. However, the study is new, and these words to describe the hard evidence are little more than a century and a half old. The *art* may be more accurately called *communication*. Archaeologists today suggest that the cave paintings, discovered in the last century, were instructional or religious or both. It would be foolish to believe that any one social group was better then or now at communicating than any other. Whatever the purposes, communication has been ignited by need.

However, let's look at some of the best communications in industrialized cultures today, likewise examples of our primordial need to convey ideas. Some examples are hardly different in many respects from those of early humans. And let's assume that art, architecture, and design are about communication, aside from their purely utilitarian goals. The best of communication may not necessarily convey messages quickly, though much of it does. The best of communication, as represented in the collection in this book, is sublimely elegant and unfolds with a narrative tied to history, like early cave art. But elegance no longer means *good taste* — an idea whose time has passed. We no longer perceive the world through the *pretty* or the *pleasing* as a primary context. An attraction for the kitsch offers some evidence. Elegance is about simplicity, the kind of simplicity, for example, expressed in the solution — by the shortest route — to a mind-twisting quantum mathematical problem.

Why is it so difficult to conjure simple solutions, solutions that capture and convey the essence of an idea? The plethora of visual garbage made, built, and realized by artists, architects, and designers smothering us in the world today provides cotemptible evidence that it is indeed difficult. The best communicators are those who know that *simple* and *simplicity* are deceptive, even misleading, words and concepts. The journey toward purity is complex, not simple. The best of the communicators must forge through the barriers of slothfulness, cowardice, and sole subservience to the marketplace. The worst of the communicators are those who surrender to slothfulness, cowardice and sole subservience to the marketplace.

All ideas in their formative states are complicated, and the methods employed by various artists, architects and designers who process ideas into reality are disparate, none favored over others. Alexander Gelman, who collected the samples here, suggests that the clearest solutions are created through a process which he has named **Subtraction**.

Even though the process of **Subtraction** is probably never easy, a deft solution may nevertheless appear to have been so, and the term **Subtraction** itself is a problematic one. When some of the artists, architects and designers find that their work has been included in these pages, they may be surprised at the label **Subtraction**, reject the term, and oppose Gelman's identification of their work as such. This rejection has occurred with *Postmodernism*. Dissension and opposition are after all autochthonous to the nature of progressive artists, architects and designers who are, thank goodness, renegades and rascals.

In an attempt to clarify the metaphor **Subtraction**, Alexander Gelman has separated the process of **Subtraction** into these dichotomies: rejection and acceptance. But the classifications are not absolute.

Using some examples from the collection here, when rejection is employed:

1. an original entity is retained, just hidden; e.g., M&Co's *5 O'clock Clock* reveals only the number 5, while the other numbers are absent.

2. an original entity is altered in a quantitative manner; e.g., for purposes of economy, cartoon hands have only three fingers and a thumb.

3. an original entity is changed qualitatively; e.g., Stefan Sagmeister's logo for a firm named Blue is printed in black on an orange background, and there is no blue color to be found.

4. or a new entity emerges when the essence of the original entity is removed; e.g., Orit Raff's photograph of the impressions on a carpet left by a piece of furniture reveals only memory.

Using other examples, when acceptance is employed:

1. an original entity is revealed through an immediate manner; e.g., MTV's logo for a television segment known as *10 Spot* is composed of only the number *10* in simple typography and a black spot.

2. an original entity is exposed with much of its depth; e.g., the basic CYMK colors (cyan, yellow, magenta, and black) used in four-color printing are shifted for the word *Gala* to offer a festive feeling.

3. an original entity is embraced without misgivings; e.g., the purely functional form of the B-2 Spirit Stealth Bomber requires — indeed, demands — no added design aesthetics.

4. or an original entity is perceived from a new point of view; e.g., Stebastian Bergne's coat-hanger substitutes the metal-wire that leaves marks and creases on clothing for the bristles and twisted wire of a bottle-cleaning brush that holds a garment in a rounded manner.

All the siblings of **Subtraction** on these pages provide an overview of these disparate approaches.

Some have been or will become known by various isms: Functionalism, Rationalism, Expressionism, Modernism, Post-modernism, Minimalism, and almost endless future addition of others. It is a quest to force styles and movements into isms, or, as the dictionary defines *isms*, into *distinctive doctrines, causes or theories*. **Subtraction** is a process, not an ism. Elegant solutions via **Subtraction** must necessarily definitely answer, *When* **Subtraction** *occurs, what has been gained?* Pure **Subtraction** never detracts from an entity but is symbiotic with an entity's internal energy.

My turgid explanation of **Subtraction** is an exercise in metaphor and simile comparable to describing the hole in a donut. In order to describe the hole (or **Subtraction**), the donut itself (or the circumstances surrounding **Subtraction**) must necessarily become the focus of the hopefully clear explanation. In other words, I have given no definition of **Subtraction** at all but rather placed it in the more capable hands of the lexicographers.

Admittedly, the categories of rejection and acceptance are pedagogical, but this book itself is not academic. No attempt has been made to draw conclusions. The approach is Aristotelian rather than Socratic; your own assumptions and scrutiny are left to roam uncorralled. Even though you will have already looked through the book before you read my explanation and may have thumbed through the pages from the back to the front, the examples have nevertheless been organized by Alexander Gelman as a narrative, for you to amble sequentially from the front to the back. Even so, you will find no suggestion that any example is better than another. By the very nature of a discrete collection like this one, the agglomeration is the result of Gelman's own process of **Subtraction**, forming an aggregation which offers some of the clearest thinking being expressed today.

The art of giving up.
by Baruch Gorkin

King Solomon's famously final statement still per-
plexes people. Is there, or isn't there anything new
under the sun, after all?

Art and all matters creative are not exempt from
this scrutiny:

Was the Post-modern architecture new in 1978?
Or was it a cynically-naive declaration that old is
new when new wants it to be?

Was the Impressionist painting new in 1873?
Or was it a revelation of the ever painterly about
painting, unleashed by the invention of photography?

Was the Modernist design new in 1919?
Or was it an eloquent pronouncement that the
Engineer was the True Artist?

Was the Renaissance sculpture new in 1435?
Or was it an Italian admission that Greeks were
cool?

It is not difficult to imagine many similar ques-
tions being asked. In words of a good friend, a computer can be programmed to ask them.

The real question is then as follows:
Is humanity's creative output defined by the externally-centric trappings of style-language, or is there something more constant and essentially creative about creativity? Not the *new* but rather the *always*.

One would expect a mega-page monograph to address such loaded a subject. Gelman, true subtractionist he, does no such thing. He skips the learned drivel and presents instead a lucid picture-thesis that effortlessly hops from one field of creative endeavour to another and in the process unveils, with a convincing starkness, an attitude common to all Creativity.

Gelman calls this attitude **Subtraction**.

Even a cursory flip through the book, loaded with content-heavy examples, is enough to realise that we are not taken back to Minimalism. **Subtraction** is equally unconcerned with both *less* and *more*.

The wide diversity of the examples Gelman chooses is also instrumental in demonstrating the suprastylistic nature of **Subtraction**.

In some works all that is insignificant is removed.
In others, the most important remains unsaid.

In some works only the artist's view matters.
In others, the artist self-removes.

In some works functionality is thwarted.
In others, functionality is all that is left.

And then there is everything in-between...

After a meditative look-read, one comes to a realization that, to Gelman, the essence of the creative process is not in the discovery of the unknown. Rather it is in the willingness to give up something known.

Perhaps this is so because the very desire to create echoes our most primal instincts: ritual sacrifice, motherhood, transference of knowledge.

Whatever the reason.

We all know it takes real courage to say something. **Subtraction** shows it takes more courage not to say something.

Forward

In **La DS** — one of his best known pieces — artist **Gabriel Orozco** removes a 25-inch wide slice along the center of a Citroen DS. By rendering this mass-produced car sleekly obsolete, he forces it into becoming an art object. The act of subtraction is the focal point of pained contemplation.

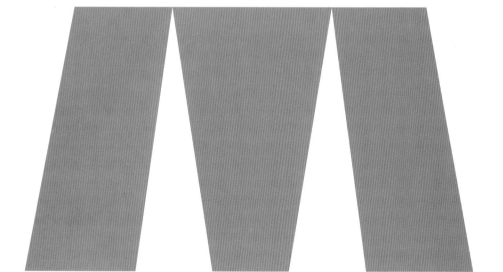

Designer **Alexander Isley** chops off the tops of
the letterforms to create a logo for **Mesa Grill**.
A slight loss of legibility is more than made
up for through the resultant direct visual
reference to a top-flat hill (mesa). The word
and the object merge.

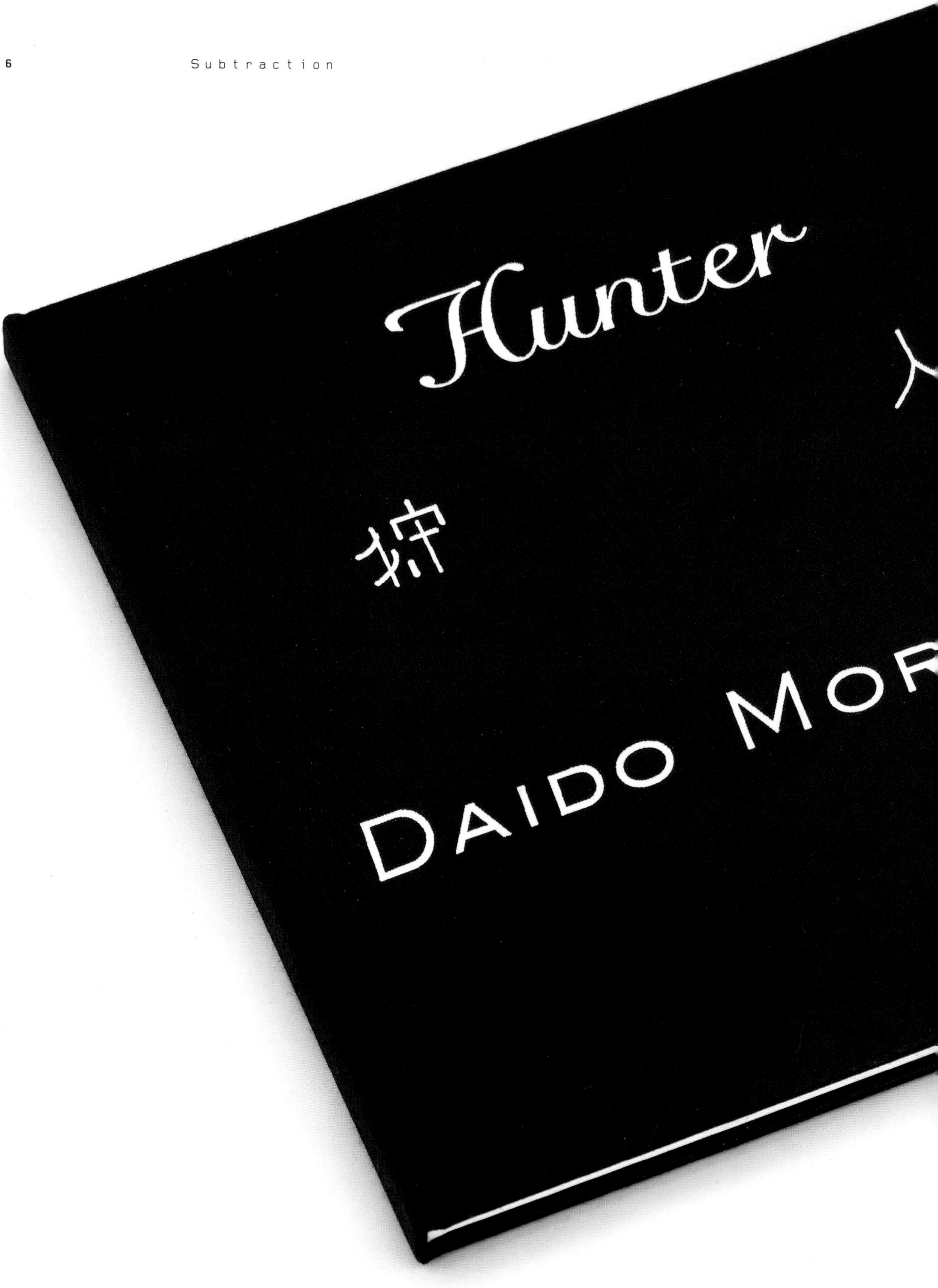
Hunter
狩
DAIDO MOR

Issay Kitagawa claims that subtraction helped sell this limited edition of Daido Moriyama's book **Hunter** in less than two weeks. By leaving the spine of the book blank, Kitagawa forced the stores to constantly display the cover.

Conventional wisdom holds that eyes play a critical role in projecting emotions. No matter, art director **Katsunori Aoki** and illustrator **Seijiro Kubo** create cartoon characters without eyes.

Their obsession became an on-going exploration into what's really essential in a facial expression.

Smile – Angry by Aoki & Kubo

Each character of the **ePhoto LCD font** designed
by **Access Factory** for the liquid crystal displays
of the **AGFA** digital cameras is confined to a
cell of 23 x 18 pixels. Instead of attempting to
mimic the roundness of letters within this very
coarse screen, the designers embrace the 45°
angularity and produce a font that is unique and
highly legible.

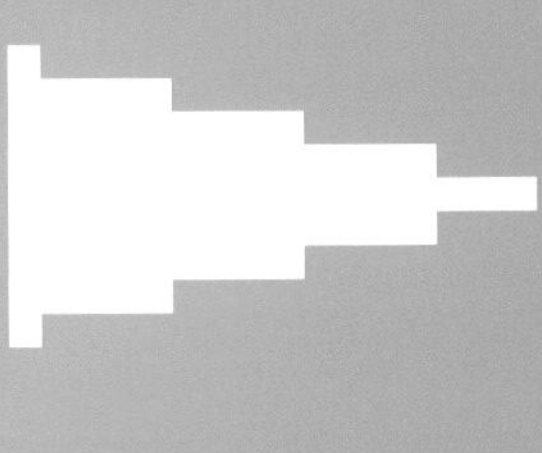

Sel

Foo

Exp

Wh

Ext

ePhoto LCD font by Access Factory

Subtraction as an experience is a frequent theme of installations by artist **Felix Gonzalez-Torres**.

The piece shown here **(Untitled)** is a neat stack of offset lithographs intended to be taken away by viewers, one at a time.

Untitled by Felix Gonzalez-Torres

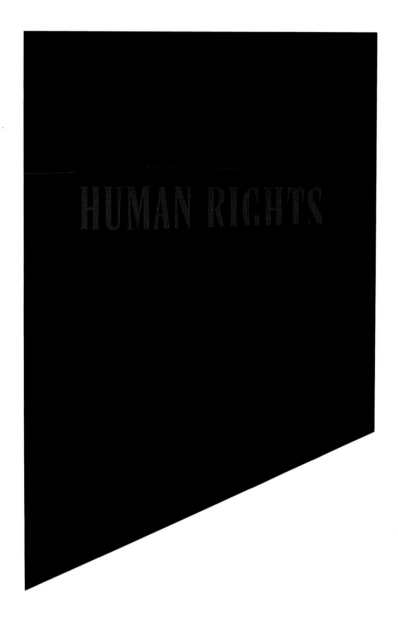

"Help take the edge off the daily reduction of Human Rights" reads the poster designed by the subtractionist **Uwe Loesch** for the 200th anniversary of the French Revolution. By "taking the edge off" the rectangular format Loesch sets off an endless loop of meanings. "The object is not in the poster, the poster itself is the object," he quips.

In the **Punktum** campaign introducing a new scanner Loesch zeroes in on a mole on a woman's face. The blemish is transformed into a literal and metaphorical expression of the slogan.

The **Uno Watch** designed by **Klaus Botta** features one hand that shows hours as well as minutes. Subtraction reduces conventional time-telling to a one step process.

Once **Nike** achieved universal brand recognition, the advertising agency **Wieden & Kennedy** removed the word "Nike" from the logo. Familiarity with the "swoosh" restores the subtracted name in our memory and a more intimate connection with the brand is forged.

400 polished stainless steel rods are arranged on a grid one mile long by one kilometer wide in Quemado, New Mexico by **Walter De Maria**. The artist has removed himself and **The Lightning Field** installation is completed by Nature. The result is always as thrillingly unruly and unpredictable as the very time, shape and intensity of the strikes.

Skeleton Key CD by Stefan Sagmeister

A carefully designed CD package for **Skeleton Key's Fantastic Spikes Through Balloon** record is punched with a grid of holes.

The random disruption of the text and images elicits an inevitably strong reaction. **Stefan Sagmeister** turns a piece of graphic design into a provocative interactive object on which an annoyed fan can clearly read the designer's wholly unpunched credit, left there for convenient blame-hanging.

Art director **Hideki Nakajima** challenges book design conventions by subtraction. His **Neatnik**, an exhibition catalog for Seiko, combines the virtues of a hard cover and a paperback.

Nakajima rejects the traditional hard cover and uses instead a single cardboard sewn into the book. The result has stiff authority but allows for flexible page flipping.

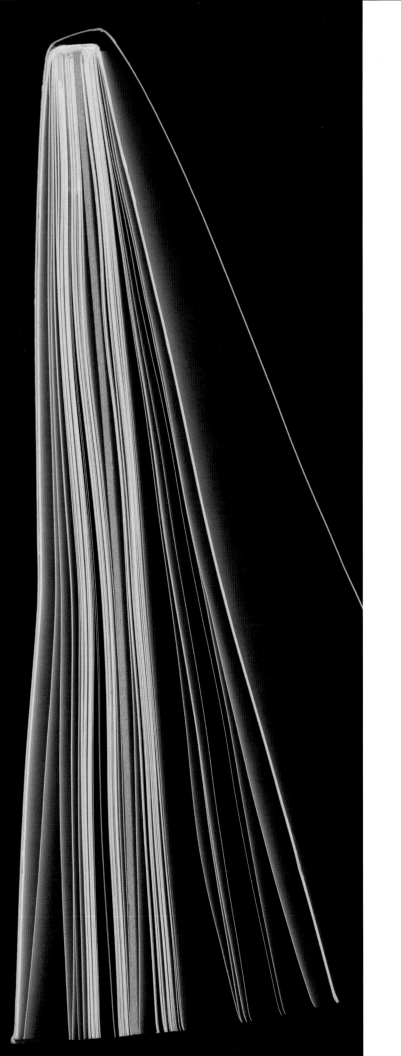

Neatnik by Hideki Nakajima

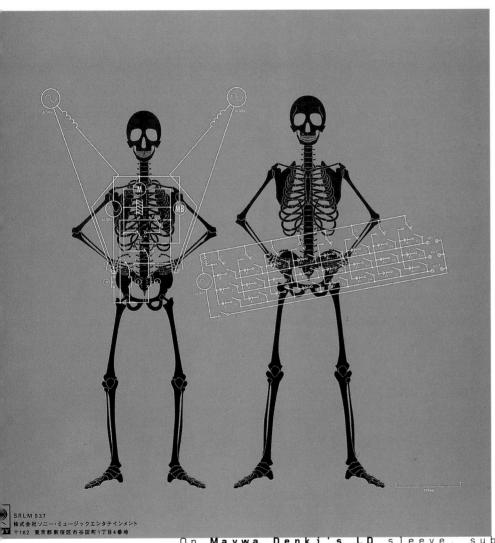

SRLM 537
株式会社ソニー・ミュージックエンタテインメント
〒162 東京都新宿区市谷田町1丁目4番地

On **Maywa Denki's LD** sleeve, subtractionist **Norio Nakamura** strips the performers and their instruments to bare bones and draws attention to the essential similarity between the human and the machine.

Ming Chueng, Lewis Lau, Tim McDowall,
Mike Tonkin and Nick Tyson of Tonkin
architects painted the entire inte[rior]
of Hong Kong movie theater Broadway
Yuen Long deep red.

Huge countdown numbers are p[rojected]
across the space to disti[nguish]
screening rooms. Peop[le]
become[s] project[ions]

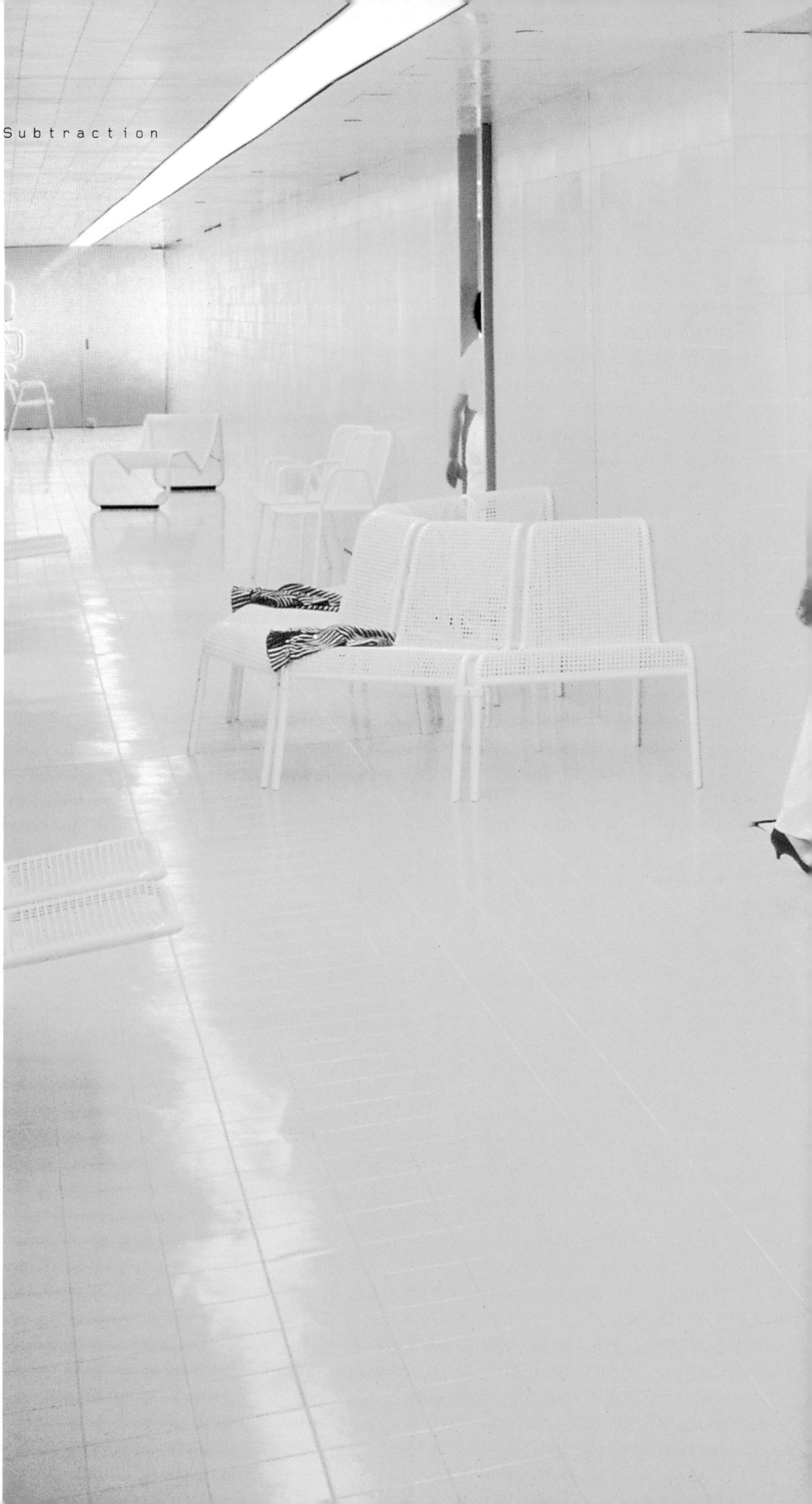

In his corporate identity for **Kroin**, designer **Massimo Vignelli** takes care of all design decisions in one shot: he paints the entire company — walls, furniture, stationery, uniforms — yellow. Since then, anything yellow is a subliminal manifestation of Kroin.

Designed by **Philippe Starck** and made by Wolford out of nylon hosiery, **StarckNaked** is a garment without a shape of its own. It is a seamless tube that can be worn as a skirt, top or a dress.

StarckNaked

Computer animated films embrace the most advanced technological possibilities: 3D effects, textures, shadows, highlights, transparency, opacity, smooth motion and other special effects.

Comedy Central's series **Dr. Katz, Professional Therapist** is instead rough, flat, and low-res. The absence of the high-tech polish accentuates the neurotic world of the show's characters.

For the 1997 **Swatch** collection, I adapted one of their classic '80s designs. By simply sliding the image off its usual place, I was able to create a totally new watch without adding or changing a single form.

In his identity for the fashion retailer **Blue**,
graphic designer **Stefan Sagmeister** subtracts
the most unexpected — the blue color.

The obvious conflict between the word and the
image fosters a stronger, more memorable brand.

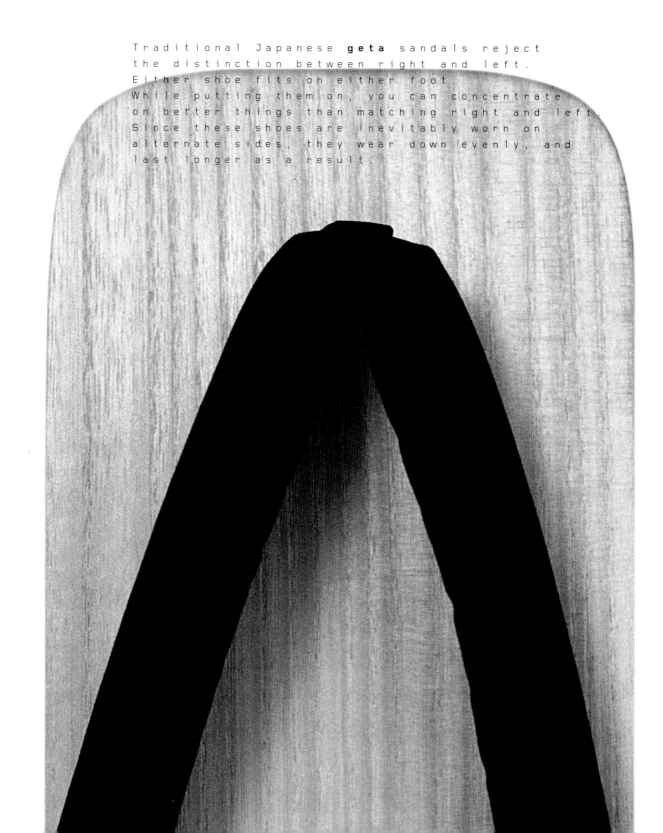

Traditional Japanese **geta** sandals reject
the distinction between right and left.
Either shoe fits on either foot.
While putting them on, you can concentrate
on better things than matching right and left.
Since these shoes are inevitably worn on
alternate sides, they wear down evenly, and
last longer as a result.

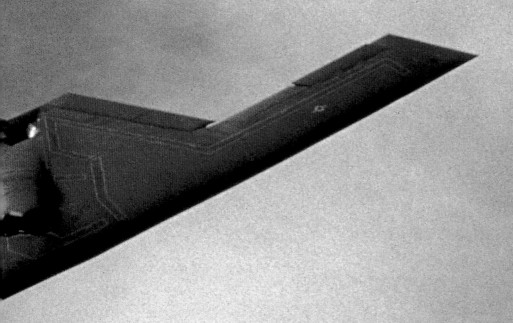

The B-2 Spirit Stealth Bomber is designed for low observability due to reduced infrared, acoustic, electromagnetic, visual, and radar signatures. Intention to please or scare did not influence its development. Despite that, it still has an eerily life-like presence.

Inspired by the simplicity and functionality of bank receipts, **Makoto Orisaki** created an identity system for **E&Y**, a Japanese furniture retailer. He employs a coarse dot-matrix printer to generate all company communication. This self-imposed restriction produces a sustained, obsessive and easy-to-maintain brand image.

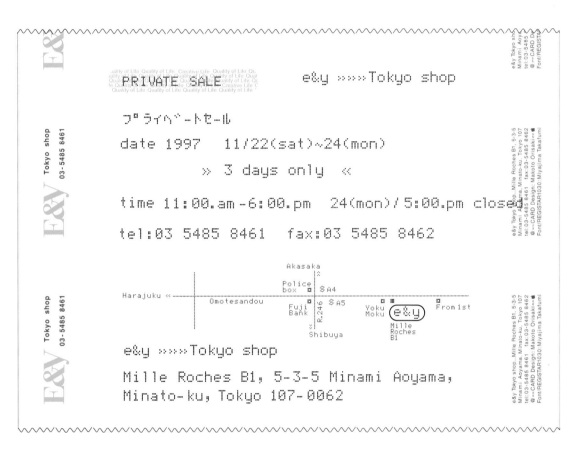

Unfinished edges and exposed seams are a frequent feature of **Comme des Garçons** collections. Designer **Rei Kawakubo** is not compelled to bring cloth to the state of completion. Subtracting the finishing steps of the garment-making process imbues her creations with an aura of uninhibited grace.

In a logo for **10 Spot** by **Tracy Boychuk** and **Jeffrey Keyton** of **MTV,** the word "spot" is subtracted and a literal spot is used instead. The result is more direct and compelling than the spelled-out name.

10
●

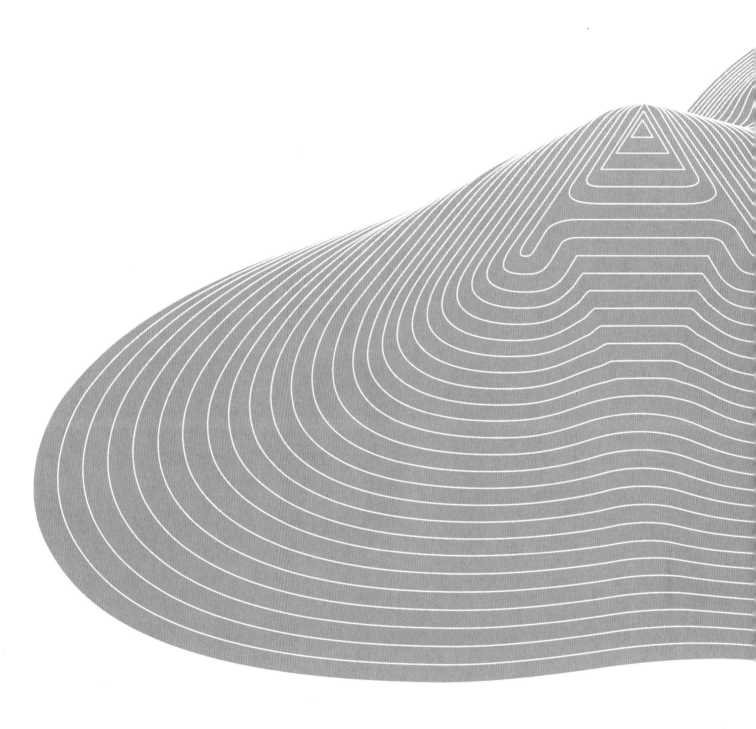

In a logo for a surveying company **ABE**, designer
Ken Miki rejects conceptual sophistication
and embraces the most obvious solution: turning
the initials of the company into a topographic
map. This relaxed simplicity yields a
surprisingly iconic mark.

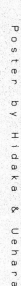

Poster by Hidaka & Uehara

In an attempt for a tongue-in-cheek promotion
of **Yasuhiro Mihara's** global ambitions, graphic
designers **Eiki Hidaka** and **Ryosuke Uehara** create
a poster that avoids all "global" imagery.

Mihara's footware creations themselves
are haphazardly thrown about and just happen
to make up the map of the world.

Droog Design discards the traditional notion of drawers as mere cavities in a single dresser.

They subtract the dresser and create a **Chest of Drawers** that is a collection of disparate old drawers encased in individual boxes and strapped with a heavy-duty luggage belt. The piece is perfectly functional and is hauntingly nostalgic.

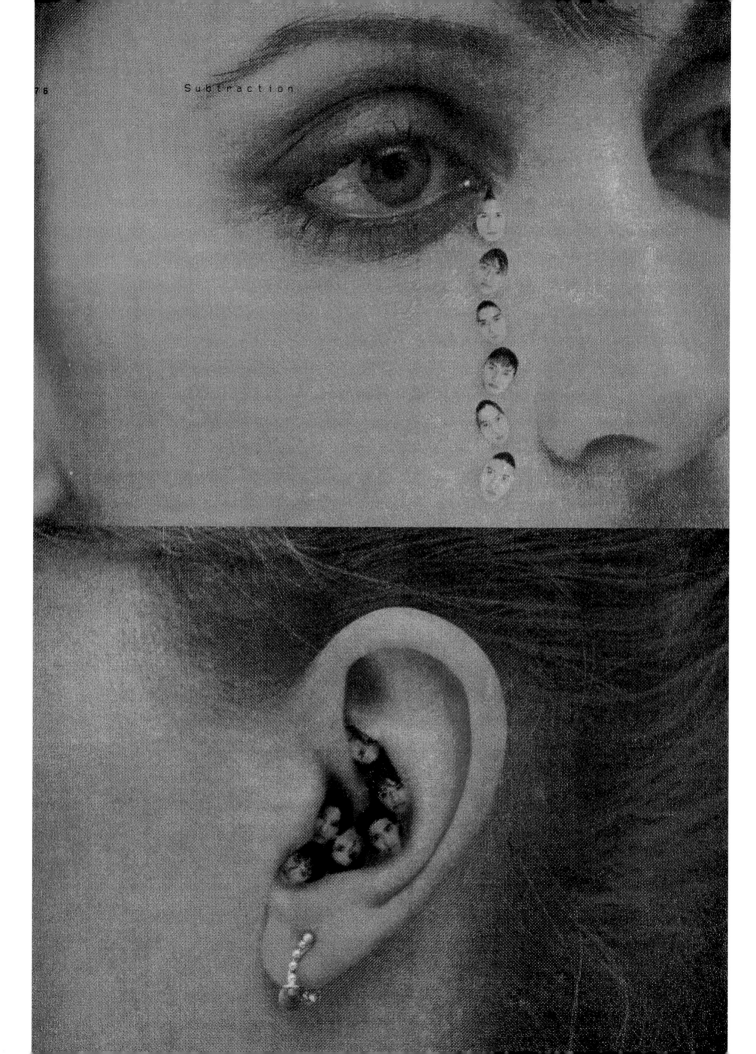

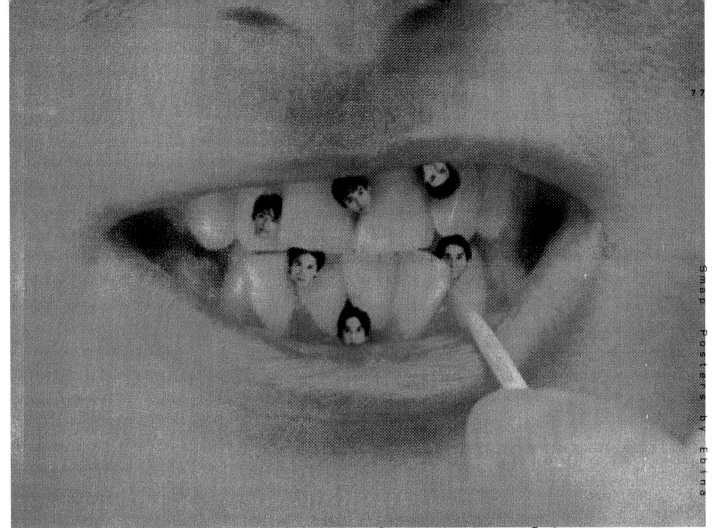

In these posters for **Smap**, designer **Tatsuo Ebina** subtracts the obligatory large photograph of the performers and even the name of the group.

By juxtaposing close-ups of a listener's face with tiny faces of the band members arranged as tears, ear plugs and particles of food stuck between the teeth, he focuses on the audience's emotional response to the music.

Instead of creating original artwork for their hand-carved rugs, **Madeleine & Dudley Edwards** faithfully reproduce famous earthworks. These images, such as **Crawley Down Crop Circle** (facing page) and **Krenkerup Labyrinth** (overleaf), coupled with the woven rug texture transcend the space of the interior and take on a confusingly monumental scale.

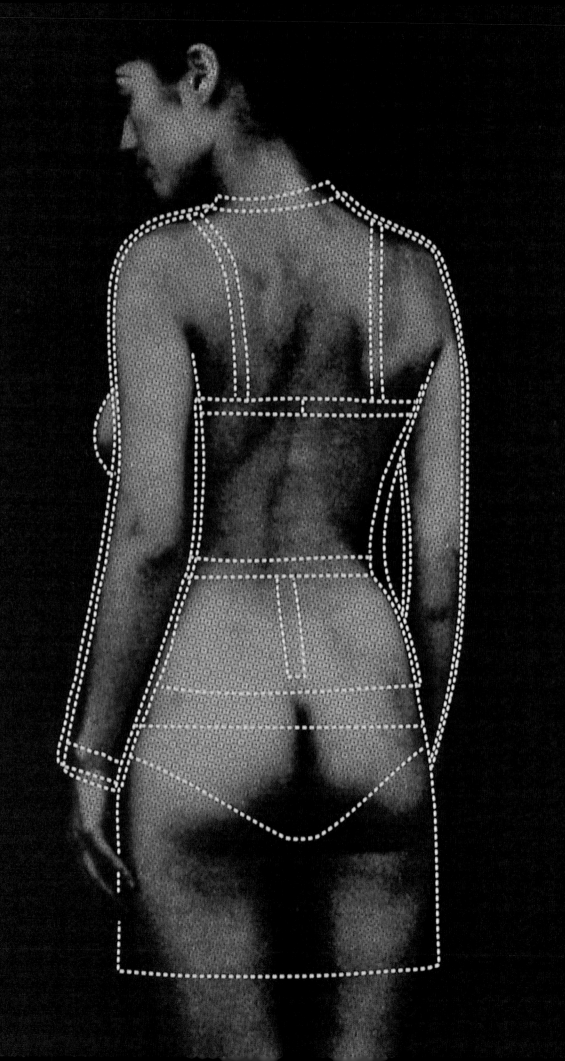

Designer **Tatsuo Ebina** subtracts everything that is not part of the human body from the posters for his 1998 one man show at the **Ginza Graphic Gallery**. By representing garments and other inanimate objects as mere dashed lines, he amplifies their subservient existence.

For his **Hole Works** series, subtractionist **Makoto Orisaki** drills neat round holes in a variety of everyday products. By diminishing his objects, such as in **#029 Day Chair** & **#030 Trash Can** (facing page), Orisaki makes them more significant through allowing the outside space to enter their inner world.

In **McDonald's "Easy"** & **Burger King "Love"** (overleaf) he goes one step further: the unpunched areas contain subliminal verbal messages.

Hole Works by Makoto Orisaki

Cloud, a serigraph by artist **Yoko Kubo**, is
a representational piece that rejects realism
and abstraction at the same time.
By walking on the very edge of recognition,
it acquires a kind of shameless mystique.

A mass-produced storage shed is retrofitted
by **Allan & Ellen Wexler** to create a **Vinyl Milford
House**, commissioned in 1994 by the **Katonah Museum
of Art**. The shed is too small to contain all the
rooms, so crate-like furniture storage cavities
are made in its walls.

Furniture is denied a permanent place in the
interior and, as a result, it moves freely between
the two-dimensional and the three-dimensional
existences.

Vinyl Mitford House by Allen & Ellen Wexler

Commissioned by the **University of Massachusetts** in 1991, the **Crate House** has only one room. Artist and architect **Allan Wexler** rethinks the very nature of living space designation. Each purpose-specific area of this house, the kitchen, the bathroom, the bedroom and the living room, fits into a crate of its own. When needed, a given crate is wheeled into an 8' x 8' x 8' room, converting it to its purpose.

For years **Christo and Jeanne-Claude** have been hiding things by wrapping them in cloth.

Through this concealment, as in **The Pont Neuf Wrapped**, Paris 1975–85, the metaphysical presence of the hidden becomes almost palpable.

Christo and Jean

Designers from the Dutch group **Lust** combine
each letter of the alphabet with 2 rectangles:
one around the letter and one inside it. By simply
subtracting various elements from that set,
they create five new typefaces that comprise their
highly versatile **Blowout** type family.

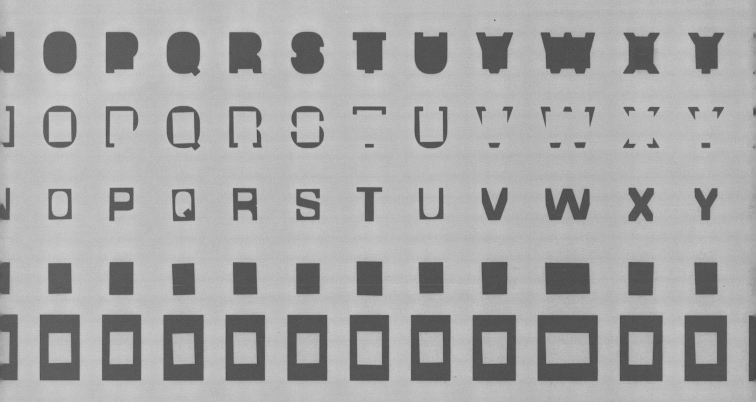

faces, food & furniture

howdy

ge... en om verschillende redenen (oa hun
vraag jij maar de rest) een fe...st op vrijdag
Mau...ice z'n nieuwe studio, de witten...stra
wear any colour, as long as it's black

Kunsthistorisches Museum - Wien

Pieter Bruegel - Het Bruiloftsmaal / Le repas de noce (1567)

Wear any color as long as it is black is a party
invitation produced by the Dutch design firm **Goodwill**
by running a Brueghel reproduction through a photocopy
machine. The witty substitution of color garments
in the famous painting with black toner removes the
seeming formality of the party dress code.

Designer **Masahiro Kakinokihara** subtracts the torso of each character that comprises his **Soccer Shirts** typeface.

The application of the numbers onto a garment worn on the torso completes the mental loop.

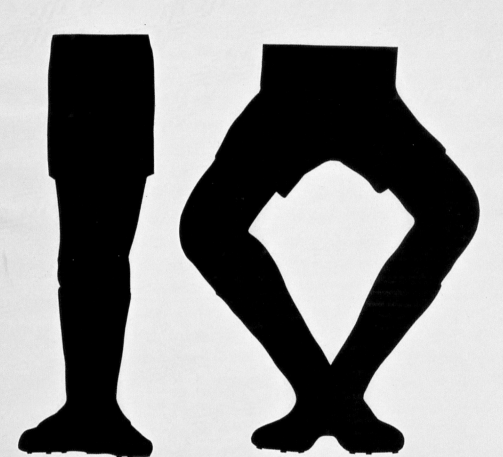

New

19

99

———

New Year's Card by Design Machine appropriates the visual language of the American supermarket.

However, only the outer shell of that abrasive vernacular is retained. The removal of the original meanings imbues the card with an understated irony.

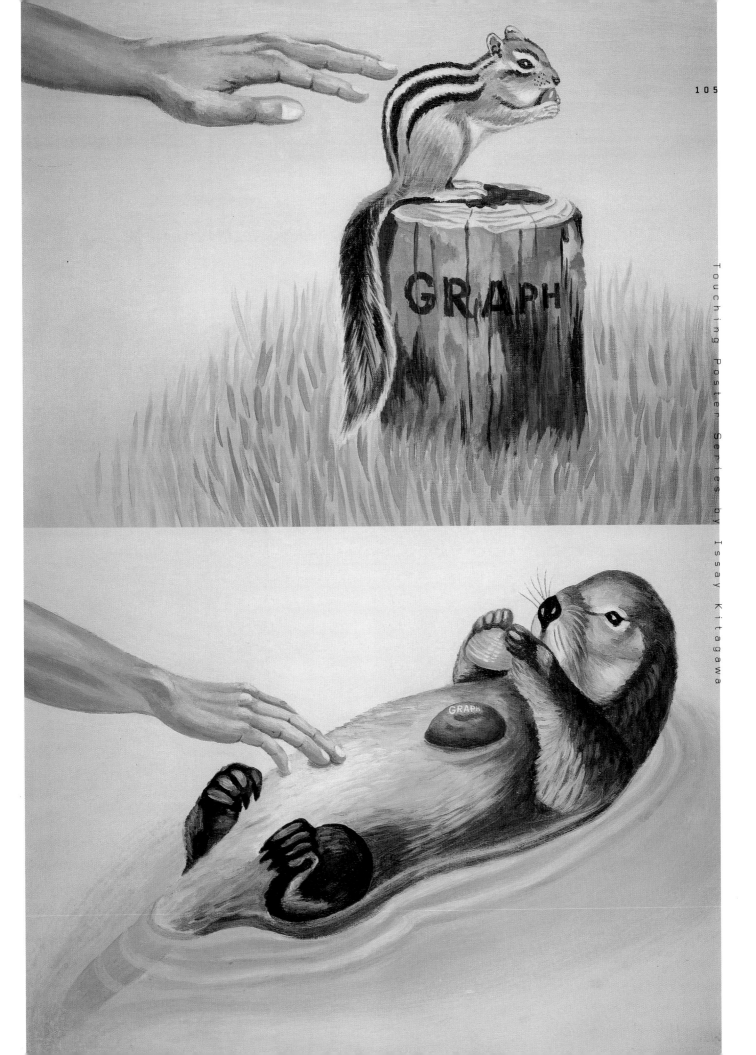

Touching Poster Series by Issay Kitagawa

Touching Poster Series by Issay Kitagawa

The Touching poster series for **Graph** by designer **Issay Kitagawa** and illustrator **Mar Sekiguchi** start off as a sequence of close-ups of a human hand patting an animal.

In the last poster, just when audience expectations are cemented, the hand is subtracted. The previously gained knowledge inevitably imprints the hand image outside of the poster and the poster's effective size far surpasses its physical dimensions.

GRAPH

M&Co.

5

5 O'Clock Clock by M&Co

The **5 O'Clock Clock** designed by **M&Co** features only one number — '5'. The absence of the other numbers makes a poignant comment on the culture of work-for-hire.

Designer **Norio Nakamura** creates his worlds out of cubes. The irreducible simplicity of the core components is what fuels an overwhelming visual complexity of his **IQ Final** video game for the **Sony Playstation**.

IQ Final by Norio Nakamura

The **Instantable** by the Scottish designer **Digby Vaughan** can be viewed as either a lamp that replaces a table or a table that replaces a lamp.

Either way, this piece, made from a single sheet of translucent polypropylene, is highly functional and far more ambient than a table with a lamp.

Instantable by Digby Vaughan

As part of his design of **Damien Hirst — The Beautiful Afterlife**, a book filled with images of cigarette butts, **Jonathan Barnbrook** creates a cover that features a postcard-like flower image.

By eliminating any connection between the cover and the content of the book, Barnbrook takes the cover out of its usual poster-wrapper role and makes it as important as the content.

A **Sun·Ad** and **Wieden & Kennedy** team consisting
of designer **Katsunori Aoki**, illustrator
Seijiro Kubo and photographer **Kouichi Ikegame**
create an outdoor advertising campaign for
Nike Japan without producing a single ad.

Instead, they supply the affiliated stores
with printed cartoon drawings & photo stickers
depicting sports star heads and Nike products.
Store clerks apply the stickers of their choice
to generate an unlimited variation of posters.

Nike Japan Posters by Katsunori Aoki

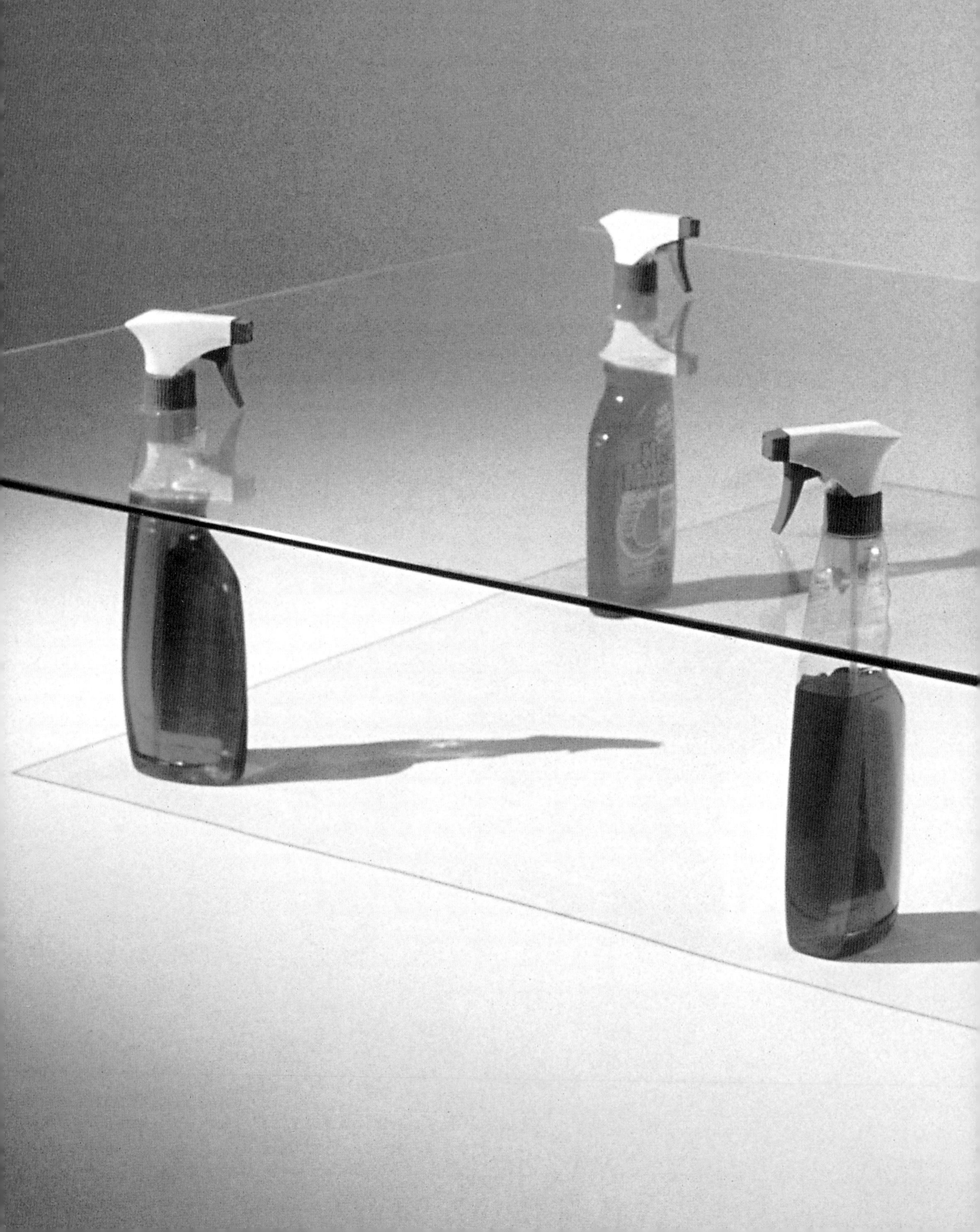

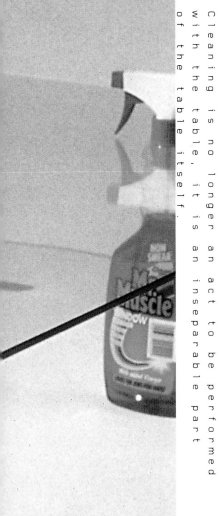

Designer **Ralph Ball** replaces the legs of his **Self-Sustaining Table** with four spray-bottles containing a generic glass cleaner.

Cleaning is no longer an act to be performed with the table, it is an inseparable part of the table itself.

Chinese calligraphy materials and methods have been
painstakingly perfected with one goal in mind:
to produce writing that will endure into posterity.

Sidewalk water calligraphy is performed on the streets
of Beijing with no such concern. This fleeting poetry
of broom and water disappears within minutes of
its creation. In the process, it offers only a pure
and concentrated experience of calligraphy.

In **Water Tower**, a 1998 installation in SoHo, New York [courtesy of Luhring Augustine and The Public Art Fund]; artist **Rachel Whiteread** fills a characteristic New York water tower with transparent resin cast and then removes the tower. A hitherto impossible experience of the inside-the-tower-space is offered.

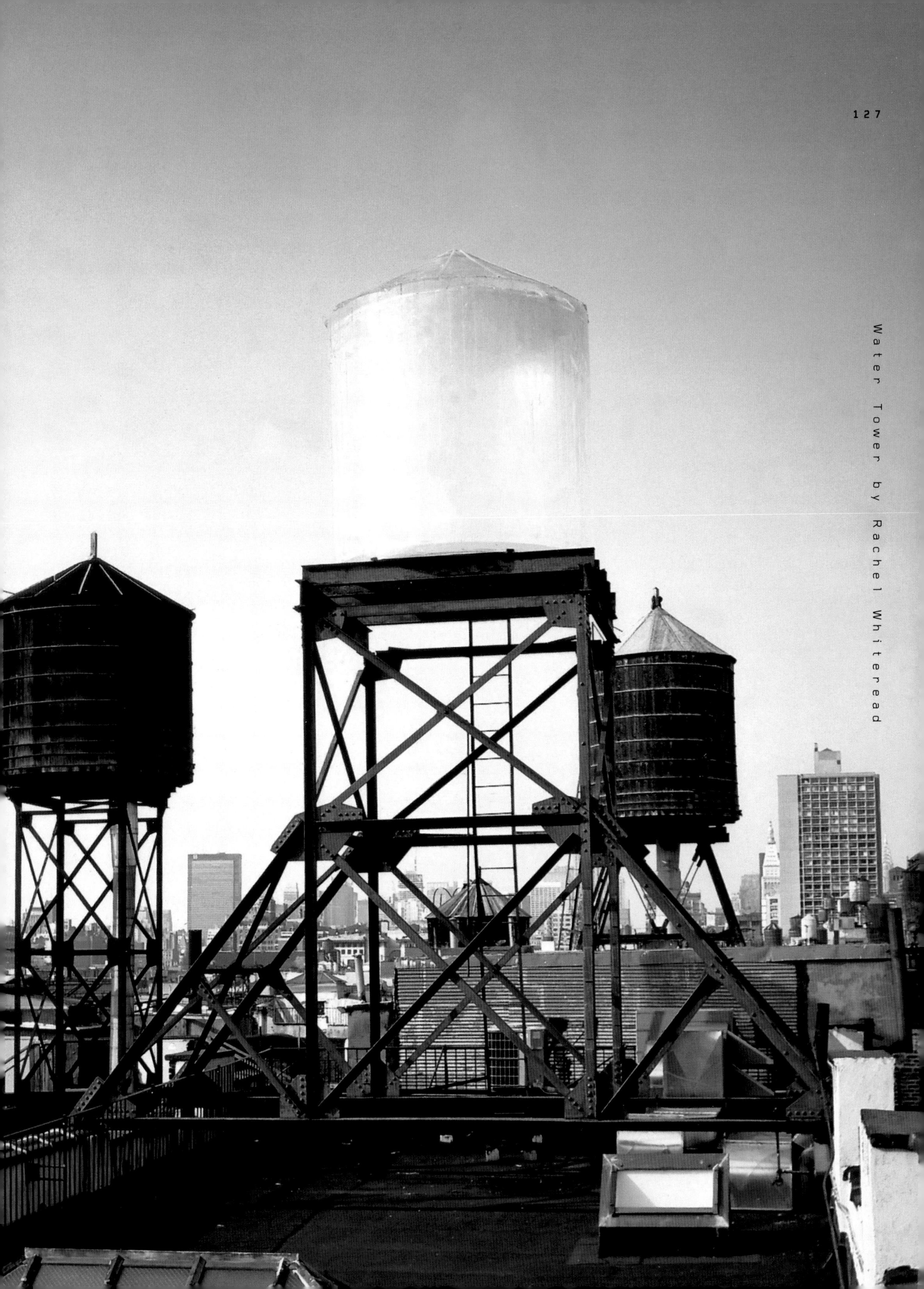

Water Tower by Rachel Whiteread

British designer **Sebastian Bergne** creates his
Mr. Mause garment hanger the same way bottle-brushes
are made: by attaching PVC bristles to a steel wire.

The resultant hanger provides a soft support for
clothes and brightly decorates an empty closet.

Mr. Mause by Sebastian Bergne

Cyan, Magenta and Yellow repetitions of the word **GALA**
shift sideways to create this "un-designed" logo for
a theatrical collaboration between the **Royal Academy** and
the **English National Opera**. Swedish designer **Barbro
Ohlson** submits to the effect produced by theatre light
gels and manages to recall it in print.

Subtraction

The London design firm **Jam** in collaboration with **Sony**
uses television monitors to create their **TV Lighting
Installation** for **IPA**. In a delicious mind twist, an
object which normally draws attention to itself —
a TV screen — takes on the role of drawing attention
to whatever it shines upon.

Mass-produced greeting cards futilely attempt to appear personalized. **Universal Greeting Cards** produced by the British firm **Daniel Eatock Design** are openly generic and honestly funny.

CASH CARD

Before giving card, insert cash/cheque and specify the amount below with a tick.

- ○ £1
- ○ £5
- ○ £10
- ○ £20
- ○ £50
- ○ £100
- ○ £200
- ○ Other*

*Specify amount £
....

↑ PLACE THIS WAY UP ON MANTLEPIECE

BIRTHDAY CARD

Before giving card, tick box or specify which birthday is being celebrated.

- ☐ Eighteenth
- ☐ Twenty-first
- ☐ Fortieth
- ☐ Fiftieth
- ☐ Hundredth
- ☐ Other*

*Please specify
....

↑ THIS WAY UP

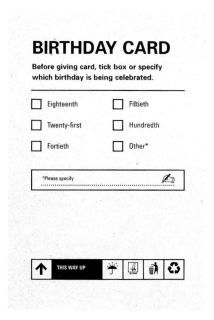

LUCKY CARD

Give somebody a chance in a million.
Affix a National Lottery Scratch Card inside.

Please note that this is a lucky card and that
any winnings must be divided equally
between:

1. Card's recipient

2. Card's sender

Recipients' and senders' names must be written in full using
capital letters in red pen. Any deviation will render the card
invalid and will result in the recipient having full rights to the
winnings.

Responsibility cannot be accepted for the
loss of this card and lottery ticket.

INSTANT CASH PRIZES TO BE WON, GOOD LUCK! ↑

LATE CARD

Write an excuse or apology in no more than
fifty words to explain why this card is late.

I promise that I will try harder next time to
make sure your card arrives on time!

Signed	Date

Inspired by Kasimir Malevich, subtractionist **John Maeda** creates his **Reactive Square**, a CD ROM featuring a black square that reacts to human voice. Depending on the selected mode, the square twists, throbs, changes color or breaks apart. The stark simplicity of the base-form makes those reactions obvious.

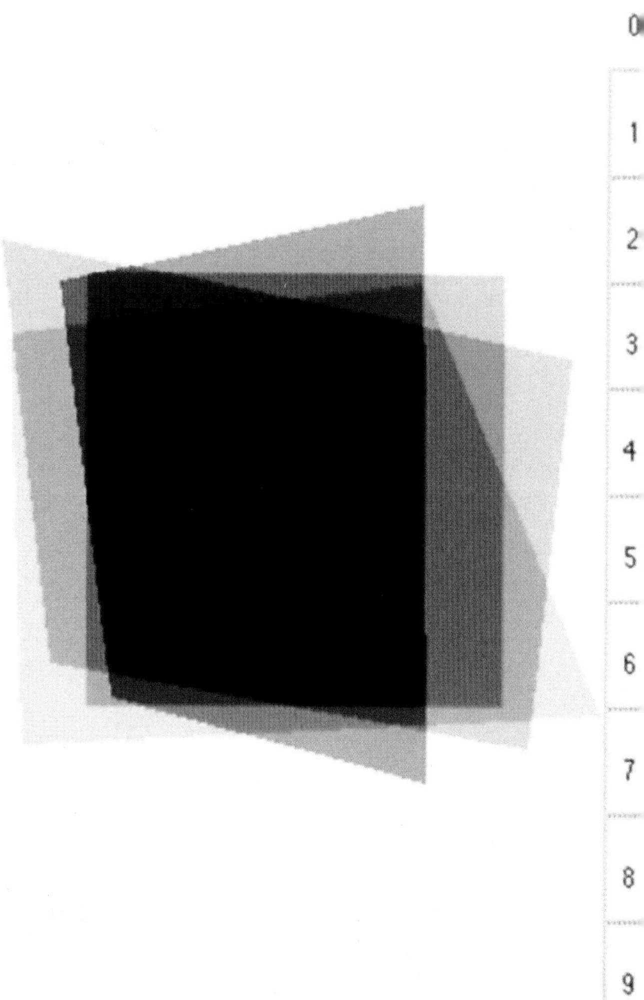
0
1
2
3
4
5
6
7
8
9

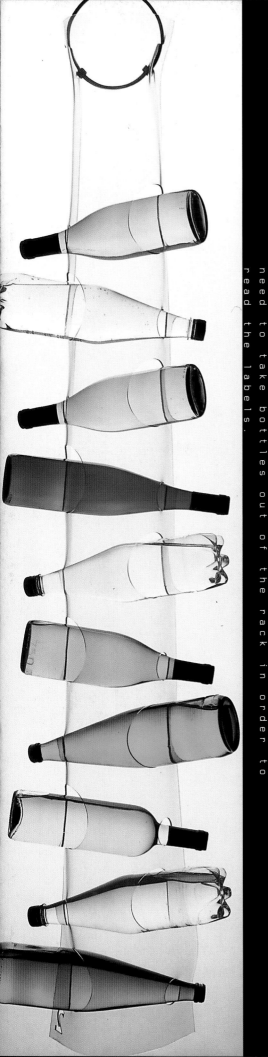

In the **Wine Rack** by **Jo Tracey** the content creates the form. The rack is nothing but a sheet of flexible transparent PVC with holes. The color, shape and personality of the rack depend entirely on the color, shape and personality of the bottles it holds.

The transparency of the rack, notes the designer, has an added functional benefit as well: there is no need to take bottles out of the rack in order to read the labels.

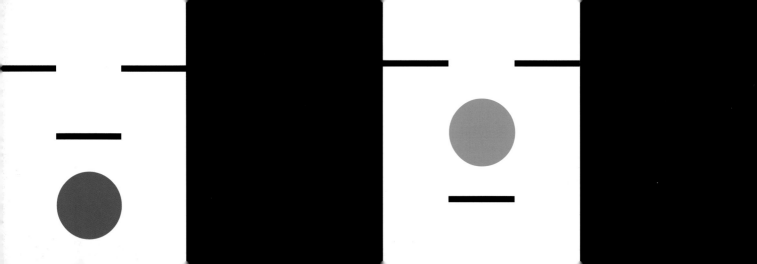

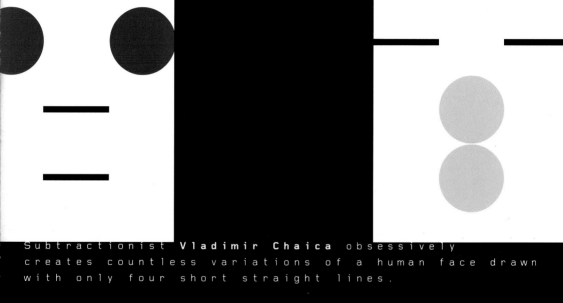

Subtractionist **Vladimir Chaica** obsessively creates countless variations of a human face drawn with only four short straight lines.

In the **Four Seasons** poster series he covers some of the lines with circles to expose the essential experience of each season.

Subtraction

Artist **Michael Jantzen** proposes to feed real-time video from cameras installed behind his beach house to LCD panels covering the entire facade of the building. An ocean-front view, normally private, is shared with others.

Artist **Orit Raff** photographs vacated dwellings.
In **Gravity and Habit** she records imprints left on the
floor surfaces by the removed furniture.
The absence of any life attributes creates a poignantly
vivid image of the elusive passage of time.

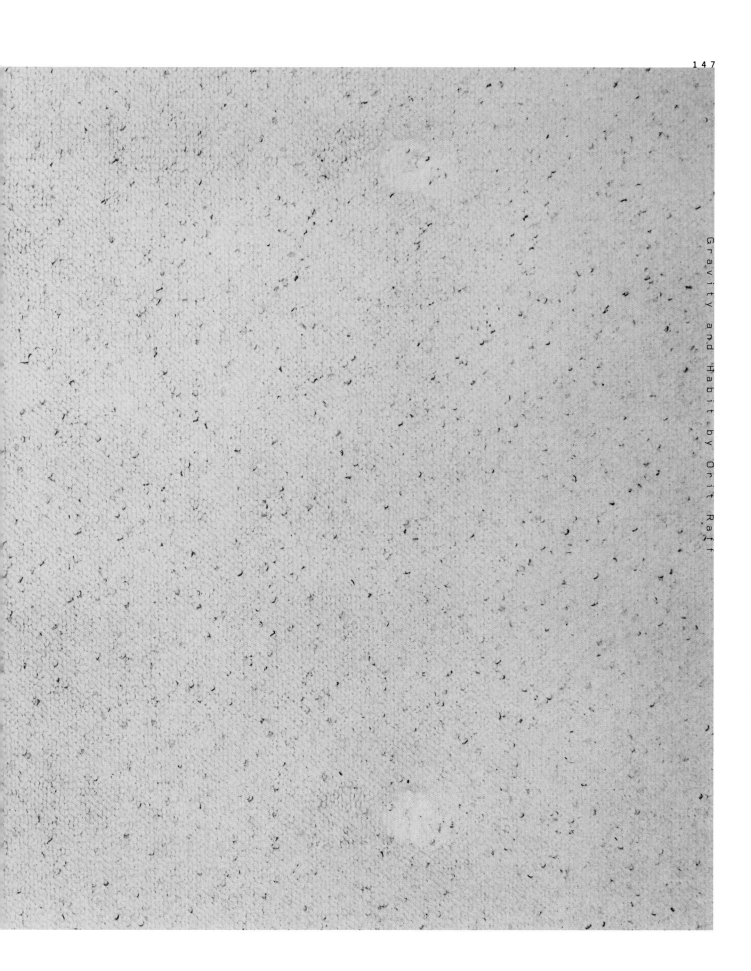

Gravity and Habit by Orit Raff

"The English buy furniture for life", says the
British designer **Matt Wingfield**.

By making his **Cardboard Coffee Table** available in a
variety of patterns, Wingfield blurs the distinction
between the table and the tablecloth. Buying a piece
of furniture is no longer a major commitment.

Table by Wingfield

In her poster series **Words In Forms**, designer
Hitomi Sago superimposes outlines of three-dimensional
wire frames over magazine title collages.

By putting these disparate elements together,
she removes their dimensional identity. The result is
a visual game: what's flat, becomes three-dimensional;
what's three-dimensional, becomes flat.

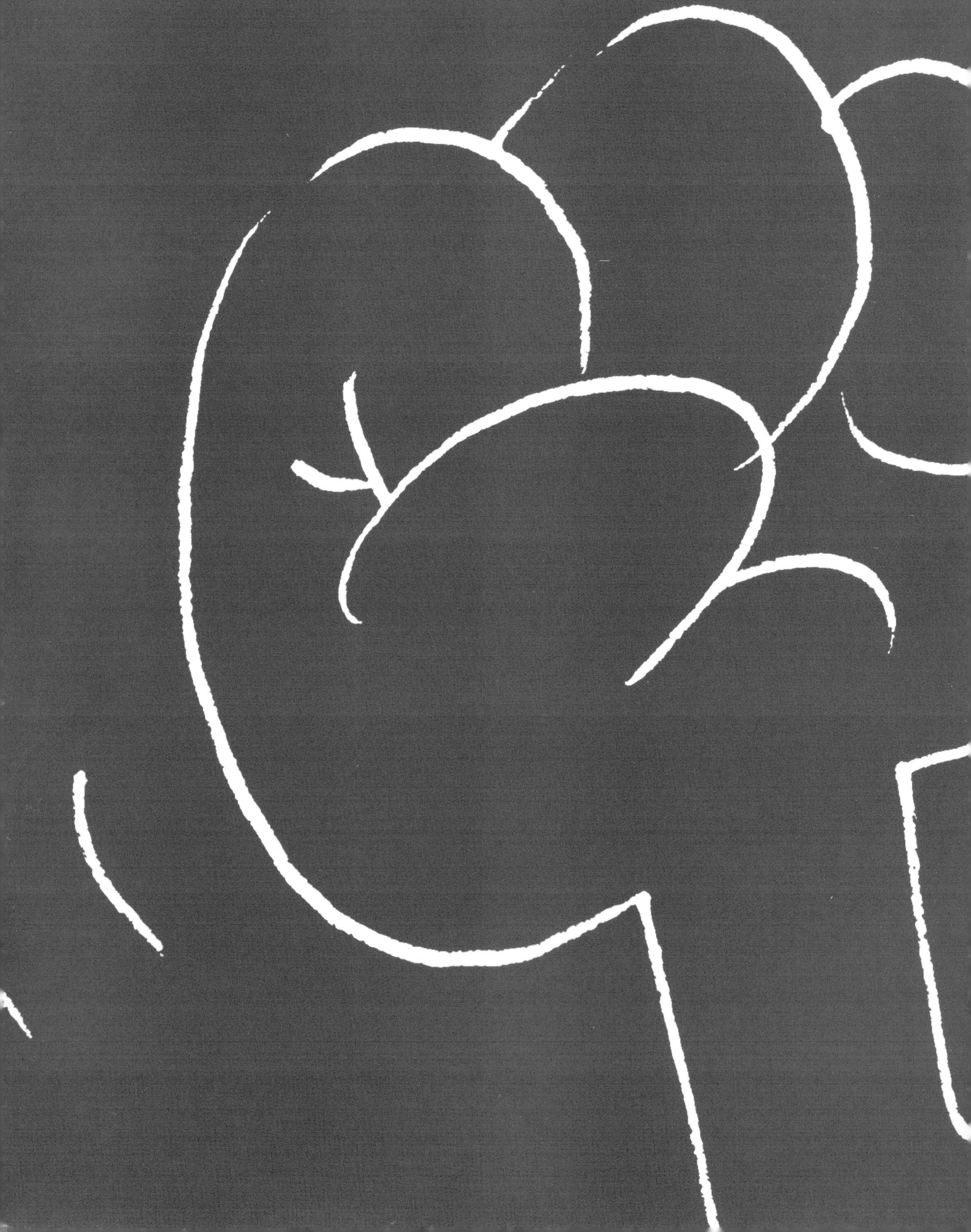

Early animators discovered that subtracting
one finger from the hands of cartoon characters
would save a great deal of time and money.
Three fingers and a thumb are now an
industry standard.

The **Camouflage** pattern is created as
a result of a process completely devoid
of any aesthetic concerns.

Camouflage confidently ignores fashion
and therein lies the secret of its
fashion allure.

Photo Credits

Images

22 Courtesy of FNAC, Paris
32 Courtesy of Andrea Rosen Gallery, New York
40 Courtesy of Dia Center for the Arts, New York
54 Courtesy of Comedy Central's Dr. Katz: Professional Therapist
56 Courtesy of Swatch Ltd.
58 Courtesy of Martin Sagmeister
62 Courtesy of U.S. Air Force
66 Courtesy of Comme des Garçons Spring/Summer collection
94 Courtesy of Christo and Jeanne-Claude
126 Courtesy of Luhring Augustine NY and the Public Art Fund
150 Courtesy of Lucy Lynn